BANTAMS

GW01454297

by

Michael and Victoria Roberts

Illustrated by

Sara Roadnight

Photographed by

Richard Roadnight

Introduction

This book is the first of our colour photographic series on domestic fowl. We have long felt the need for an accurate and modern colour guide and the text has been kept to a minimum in favour of the photographs. Should you require the precise standards and show points then this book should be read in conjunction with *The British Poultry Standards* (Butterworth Press). All the birds photographed are show winners, so can be taken to be representative of their breed, having been judged under Poultry Club rules. Most of the popular and rare breeds are displayed.

Bantams in Colour will enable the reader to identify most breeds of bantams, to produce birds conforming to the correct standard and to realise the enormous diversity of colour and type available. Some top breeders are producing new colour variations within the breeds, not all of which we have been able to include. There are original paintings by Sara Roadnight of feather types and markings to aid identification and conformity, a diagram of the various parts of a chicken with the technical names, plus egg colour variation.

A true bantam is one which does not have a large fowl counterpart, so the majority of birds represented here are technically miniatures. Incidentally, the name bantam is said to have originated in Indonesia where there is a town on the north-west end of Java called Bantam. The connection is uncertain but may have involved small poultry from this area.

We would like to thank the members of the Committee of the National Federation of Poultry Clubs and the Poultry Club of Great Britain for allowing us to photograph the birds and we are indebted to our photographer, Richard Roadnight, for his patience and skill.

Michael and Victoria Roberts
The Domestic Fowl Trust

Contents

Inside back cover: Labelled diagram of a chicken showing characteristics of most breeds.

Ancona, single comb—M. Cropper, Preston, Lancs.

Ancona, single comb—T. Heginbotham, Preston, Lancs.

Breed	Ancona
Type	Miniature
Classification	Light
Origin	Italy
Colour variations	Mottled
Egg colour	White
Specifications	Upright bearing, single or rose comb, white ear lobes, legs yellow but mottled with black. If single comb, in the male it should be erect and in the female it should flop over on one side.
	Excellent egg layer, but sometimes nervous.
Faults to avoid	Over size.

Rumpless Araucana—Miss A. L. Clark, Hockley Heath, Warks.

Lavender Araucana—G. Hamer & Son, Bolton, Lancs.

Breed	Araucana
Type	Miniature
Classification	Light
Origin	Chile and Peru
Colour variations	Lavender, black, brown, any Old English Game colour
Egg colour	Blue
Specifications	This breed also has a rumpless variation. Legs are dark grey to black. Pea comb, very small in the female. There should be a small crest, not upright, muffing. If rumpless there is no crest but long ear tufts.

Good egg layer, hardy. |
| **Faults to avoid** | Oversize, lack of muffing |

Australorp—R. E. Sharpe & Son, Preston, Lancs.

Australorp—L. Hattam, Redruth, Cornwall.

Breed	Australorp
Type	Miniature
Classification	Heavy
Origin	England and Australia
Colour variations	Black
Egg colour	Cream to light brown
Specifications	A stocky breed with single comb and black legs. Red ear lobes. Compact but upright tail. Good egg layer.
Faults to avoid	Long flowing tail.

Black Mottled Barbu d'Uccle—Miss C. Chadwick, Reading, Berks.

Millefleur Barbu d'Uccle—A. J. Davies, Haverfordwest, Dyfed.

Breed	Barbu d'Uccle
Type	Bantam
Origin	Belgium
Colour variations	Cuckoo, black mottled, black, blue quail, lavender lavender mottled, laced-blue, millefleur, porcelaine, white, quail
Egg colour	Cream
Specifications	Short, pert bearing, single comb, heavily feathered legs and feet. Also muffs and beard.
Faults to avoid	Incorrect comb and poorly feathered legs and feet.

Black Barbu d'Anvers—M. A. Allen, Sutton Coldfield, West Midlands.

Quail Barbu d'Anvers—F. R. Simmonds, Weymouth, Dorset.

Breed	Barbu d'Anvers
Type	Bantam
Origin	Belgium
Colour variations	Cuckoo, black mottled, black, quail, blue quail, lavender, lavender mottled, laced-blue, millefleur, porcelaine, white
Egg colour	Cream
Specifications	Upright, rose comb and clean legs. Wings held nearly vertically in the male. Bearded and muffed.
Faults to avoid	Single comb, large wattles and ear lobes, feathered legs.

Dark Brahma—R. Parsons, Devizes, Wilts.

Gold Brahma—J. Smith, Banbury, Oxon.

Gold Brahma—H. F. Cameron, Kettering, Northants.

Breed	Brahma
Type	Miniature
Classification	Heavy
Origin	India
Colour variations	Light, dark, white, gold, birchen
Egg colour	Cream to light brown
Specifications	Upright, stocky and strong front, pea comb, feathered legs and feet, legs yellow in colour.
Faults to avoid	Too large a tail and legs too long. White ear lobes, clean legs.

Light Brahma—Mrs. V. Marshall, Bolton, Lancs.

Birchen Brahma—J. E. Buck, Bristol.

Light Brahma—M. Hatcher, Newbury, Berks.

Andalusian—D. H. Critchlow, Stoke-on-Trent, Staffs.

Andalusian—Mrs. F. H. Dowden, Sandy, Beds.

Breed	Andalusian
Type	Miniature
Classification	Light
Origin	Spain
Colour variations	Blue
Egg colour	White
Specifications	Upright bearing, single comb, white ear lobes. In the male the comb should be well serrated and erect and in the female it should flop over. Legs dark grey to black. Feathers blue with black lacing on body.
	Excellent egg layer, can be flighty.
Faults to avoid	Lack of lacing.

Black Croad Langshan—R. D. Gilbert-Cunliffe, Warlingham, Surrey.

Black Croad Langshan—Miss J. Murch, Barnstaple, North Devon.

Breed	Black Croad Langshan
Type	Miniature
Classification	Heavy
Origin	China
Colour variations	Black, white rarely
Egg colour	Light brown
Specifications	Single comb, lightly feathered legs and feet, greeny tinge to black plumage, white to pink legs. Proud carriage.
Faults to avoid	White ear lobes, no feathers or too large feathers on legs, yellow legs.

Silver Grey Dorking—R. C. Tucker, Braunton, North Devon.

Silver Grey Dorking—D. Hoyle, Rossendale, Lancs.

Breed	Dorking
Type	Miniature
Classification	Heavy
Origin	England
Colour variations	Silver grey, red, white, cuckoo
Egg colour	Cream
Specifications	Single comb in red and silver grey, rose comb in cuckoo and white. Five toes. Boat-shaped body, regal carriage. White legs and feet. Wings horizontal with the ground and long saddle hackles.
Faults to avoid	Four-toed birds, incorrect comb, upright and short body.

Black Faverolle—Mrs. M. Appleby, Barton-under-Needwood, Staffs.

Ermine Faverolle—J. Kraft, Hunmanby, North Yorks.

Salmon Faverolle—Mrs. S. Bruton, Wolverhampton.

Salmon Faverolle—Mrs. S. Bruton, Wolverhampton.

Breed	Faverolle
Type	Miniature
Classification	Heavy
Origin	France
Colour variations	Black, salmon, ermine, blue, buff, white
Egg colour	Cream
Specifications	Single comb, upright bearing, five toes, beard and side muffs, lightly feathered legs. Red ear lobes.
Faults to avoid	Lack of fifth toe, clean legs, no beard or muffing.

White Frizzle—W. H. Henderson, Winterton, South Humberside.

Buff Frizzle—Mrs. P. Burdett, Thirsk, North Yorks.

Black Frizzle—Mrs. E. Amos, Llanwrtyd Wells, Powys.

Breed	Frizzle
Type	Miniature
Classification	Light
Origin	China
Colour variations	Black, white, buff, blue, silver grey
Egg colour	Cream
Specifications	Each feather curling towards the head of the bird. Single comb, red ear lobes, legs yellow.
Faults to avoid	Lack of even curl in feathers, long tail.

Indian Game—H. R. Jones, Kilkhampton, Cornwall.

Jubilee Indian Game—J. Sherratt, Wem, Shropshire.

Breed	Indian Game (Cornish)
Type	Miniature
Classification	Heavy
Origin	Cornwall, England
Colour variations	Indian, jubilee
Egg colour	Cream
Specifications	Indian colour is chestnut, laced with greeny black. Jubilee colour is chestnut, laced with white. Very stocky and broad with thick yellow legs set well apart. Meaty looking. Pea comb and red ear lobes.
Faults to avoid	Incorrect comb, leggy, narrow-chested.

Gold Spangled Hamburg—
R. & T. Smitham, Camborne, Cornwall.

Silver Spangled Hamburg—G. Macnab, Knutsford, Cheshire.

Silver Spangled Hamburg—A. R. Buckland, Ascot, Berks.

Breed	Hamburg
Type	Miniature
Classification	Light
Origin	Europe
Colour variations	Gold pencilled, gold spangled, silver pencilled, silver spangled
Egg colour	White
Specifications	Rose comb, white ear lobes, upright and alert. Fine, elegant bearing.
	Black or white bantam Hamburgs are not recognised as separate colours due to the similarity to Rosecombs.
Faults to avoid	Incorrect comb.

Gold Pencilled Hamburg—
D. H. Critchlow, Stoke-on-Trent, Staffs.

Gold Pencilled Hamburg—A. J. Spencer, Hailsham, Sussex.

Silver Pencilled Hamburg—J. D. Owen, Camberley, Surrey.

Grey Japanese—G. Turner, Sandbach, Cheshire.

Black Japanese—T. V. Richards, Redruth, Cornwall.

Black-tailed White Japanese—Dr. A. N. Lockhart, Dunstable, Beds.

Mottled Japanese—G. Turner, Sandbach, Cheshire.

Breed	Japanese
Type	Bantam
Origin	Japan
Colour variations	Black, blue, mottled, black-tailed white, grey, buff, cuckoo, black-tailed buff
Egg colour	Cream
Specifications	Very short, clean legs, single comb, long upright tail, wings carried low, red ear lobes.
Faults to avoid	Long legs, low and short tail.

White Leghorn—I. W. Sissons, Pocklington, York.

Brown Leghorn—R. Grice, Darlington, Co. Durham.

Barred Leghorn—R. D. Gilbert-Cunliffe, Warlingham, Surrey.

Black Leghorn—C. Lowe, Sheffield, South Yorks.

Breed	Leghorn
Type	Miniature
Classification	Light
Origin	Italy
Colour variations	White, blue, black, brown, cuckoo, duckwing
Egg colour	White
Specifications	Tall breed, large single comb in male, female's comb must flop over. White ear lobes, yellow legs. Rose comb sometimes seen.
	Good egg layer.
Faults to avoid	Incorrect leg colour, erect comb in female.

Marans—T. Moss, Nantwich, Cheshire.

Marans—Mrs. D. Pickett, Chichester, West Sussex.

Breed	Marans
Type	Miniature
Classification	Heavy
Origin	Holland
Colour variations	Cuckoo
Egg colour	Dark brown
Specifications	Single comb, red ear lobes, white legs, stocky build.
	The name Marans was given to the chickens brought by Dutch drainage engineers to the region of Marans in France in the sixteenth century.
Faults to avoid	Yellow legs, white ear lobes, pale eggs.

Black Minorca—R. & E. Shaw, Dewsbury, West Yorks.

Black Minorca—R. & E. Shaw, Dewsbury, West Yorks.

Breed	Minorca
Type	Miniature
Classification	Light
Origin	Spain
Colour variations	Black, white, buff, blue
Egg colour	White
Specifications	Tall, upright bearing, large single comb flops over in the female, large white ear lobes, black legs. White legs only in white colour variation. Good egg layer. Eggs large.
Faults to avoid	Yellow legs, small ear lobes, erect comb in female.

Black Malay—D. A. Scrivener.

White Malay—D. A. Scrivener.

White Malay—D. A. Scrivener.

Breed	Malay
Type	Miniature
Classification	Heavy
Origin	Asia
Colour variations	Black, white, spangled, pile, duckwing
Egg colour	Cream
Specifications	Very tall, leggy, wiry bird, walnut comb. Yellow legs and red ear lobes.
Faults to avoid	Lack of height, stockiness, incorrect comb.

22

Pile Modern Game—I. Eva, Redruth, Cornwall.

Birchen Modern Game—M. Hadfield, Disley, Cheshire.

Brown-red Modern Game—J. A. Richardson, Southampton, Hants.

Silver Duckwing Modern Game—
J. Hague, Stockport, Cheshire.

Breed	Modern Game
Type	Miniature
Classification	Heavy
Origin	England
Colour variations	Pile, birchen, brown-red, black-red, silver duckwing, gold duckwing, black, blue, white
Egg colour	Cream
Specifications	Small single comb, but customary to dub the male combs. Very long legs, tall, well-rounded muscular bird.
Faults to avoid	Lack of height, incorrect comb. Check leg colour matches corresponding colour variation.

Nankin—K. J. Bosley, Wantage, Oxon.

Nankin—F. & J. Gidley, Flitwick, Beds.

Breed	Nankin
Type	Bantam
Origin	China
Colour variations	Ginger-buff
Egg colour	Cream
Specifications	Single comb or rose comb, red ear lobes, white legs, male should be orange marmalade colour, hen slightly paler.
Faults to avoid	Lack of black main tail feathers.

Silver Duckwing Old Dutch—B. Short, Westbury, Wilts.

Breed	Old Dutch
Type	Bantam
Origin	Holland
Colour variations	Silver duckwing, gold duckwing, salmon, blue duckwing, partridge
Egg colour	Cream
Specifications	Single comb, upright, pert bearing, legs slate coloured, white ear lobes, wings held vertically, full tail, short back.
Faults to avoid	Red ear lobes, yellow legs, narrow build.

Pile Old English Game—T. & M. Dunn, Newcastle, Staffs.

Crele Old English Game—Paradise Prints, Thirsk, North Yorks.

Silver Duckwing Old English Game—
Mrs. B. Palmer, Clitheroe, Lancs.

Silver Duckwing Old English Game—
M. A. Allen, Four Oaks, West Midlands.

Breed	Old English Game
Type	Miniature
Classification	Light
Origin	England
Colour variations	Pile, crele, silver duckwing, wheaten, spangle, partridge, black-red, blue dun, white, black, furnace, cuckoo, ginger-red, henny, yellow duckwing
Egg colour	Cream
Specifications	Small single comb, but customary to dub male combs. Well-rounded, neat muscular bird, heart-shaped when viewed from above.
Faults to avoid	Over size. Check leg colour matches corresponding colour variation.

Wheaten Old English Game—Evans Bros, Evesham, Worcs.

Spangle Old English Game—
D. J. Lowe & Sons, Middlewich, Cheshire.

Partridge Old English Game—E. J. White, Stadhampton, Oxon.

Black-red Old English Game—M. J. Woolway, Swansea.

White Orloff—N. G. Hughes, Leicester.

White Orloff—N. G. Hughes, Leicester.

Breed	Orloff
Type	Miniature
Classification	Heavy
Origin	Russia
Colour variations	White, black, mahogany, spangled
Egg colour	Cream
Specifications	Pea comb, yellow legs, upright bearing, thick neck, solid looking, ear lobes red, small muff and beard.
Faults to avoid	Incorrect leg or comb, lack of muffs and beard.

Buff Orpington—Mrs. R. J. Lewis, Lingfield, Surrey.

Blue Orpington—D. Bruce, Great Missenden, Bucks.

Buff Orpington—Mrs. R. J. Lewis, Lingfield, Surrey.

Black Orpington—Dr. W. C. Carefoot, Preston, Lancs.

White Orpington—A. J. Robinson, Foston, Derbyshire.

Breed	Orpington
Type	Miniature
Classification	Heavy
Origin	Kent, England
Colour variations	Buff, blue, white, black, cuckoo
Egg colour	Light brown
Specifications	Compact shape, small single comb, red ear lobes, white legs, feathers soft and fluffy. Black variety may have rose comb.
Faults to avoid	Over size single comb, legginess.

Buff Pekin—P. Hinton, Knockholt, Kent.

Cuckoo Pekin—J. D. Hoyland, Barnsley, South Yorks.

Blue Pekin—J. D. Hoyland, Barnsley, South Yorks.

White Pekin—J. C. McNeill, Ferndown, Dorset.

Mottled Pekin—T. L. Corner, Sheffield, South Yorks.

Breed	Pekin
Type	Miniature (of Cochin)
Classification	Heavy
Origin	China
Colour variations	Black, white, buff, cuckoo, mottled, blue, lavender, partridge, barred, columbian
Egg colour	Cream
Specifications	Single comb, red ear lobes. Should be circular in shape when viewed from above, whole outline rounded, heavily feathered legs and feet.
Faults to avoid	Over size, lack of feathers on feet.

Gold Duckwing Phoenix—D. Hoyle, Rossendale, Lancs.

Gold Duckwing Phoenix—D. Hoyle, Rossendale, Lancs.

Breed	Phoenix
Type	Miniature
Classification	Light
Origin	Japan
Colour variations	Silver duckwing, gold duckwing
Egg colour	Cream
Specifications	Single comb, tail must be at least twice length of body in male, female has longish tail held horizontally. Legs yellow.
Faults to avoid	Pea comb. Beware confusion with Yokohama, a rare breed. In Britain the Phoenix is sometimes mistakenly called Yokohama.

Gold Poland—W. H. Henderson, Winterton, South Humberside.

Chamois Poland—S. Robinson, Liskeard, Cornwall.

White-crested Black Poland—G. S. Martin, Solihull, West Midlands.

Silver Poland—D. Patrick, Stamford, Lincs.

Breed	Poland
Type	Miniature
Classification	Light
Origin	Europe
Colour variations	White-crested black, black, white, gold, silver, white-crested blue, chamois
Egg colour	White
Specifications	Upright with profuse crest, compact in female, more open and spiky in male, uniform in colour. Small horn comb, white ear lobes.
Faults to avoid	Lack of crest, uneven shape and colour of crest. Check leg colour matches corresponding colour variation.

New Hampshire Red—Mr. & Mrs. F. Cherry, Great Missenden, Bucks.

New Hampshire Red—Mr. & Mrs. F. Cherry, Great Missenden, Bucks.

Breed	New Hampshire Red
Type	Miniature
Classification	Heavy
Origin	U.S.A.
Colour variations	Chestnut
Egg colour	Light brown
Specifications	Single comb, red ear lobes, yellow legs. Black-edged wings, main tail feathers black. Upright bearing, solid and broad.
Faults to avoid	Incorrect leg colour or comb, lack of black feathers.

Wheaten Rumpless Game—P. H. Rains, Weston Underwood, Derby.

Pile Rumpless Game—I. T. Rollason, Penryn, Cornwall.

Breed	Rumpless Game
Type	Miniature
Classification	Light
Origin	England
Colour variations	As Old English Game
Egg colour	Cream
Specifications	Single comb, distinguishing feature of no parson's nose and no tail, otherwise similar to Old English Game, but not dubbed.
Faults to avoid	Over size.

Rhode Island Red—M. Ashman, Frome, Somerset.

Rhode Island Red—M. Ashman, Frome, Somerset.

Breed	Rhode Island Red
Type	Miniature
Classification	Heavy
Origin	U.S.A.
Colour variations	Dark mahogany with beetle-green sheen
Egg colour	Light brown
Specifications	Single or rose comb, yellow legs, red ear lobes, solid carriage. Black-edged wings, main tail feathers black.
	Good egg layer.
Faults to avoid	Incorrect leg colour, white ear lobes, light coloured plumage.

Barred Plymouth Rock—Dr. W. C. Carefoot, Preston, Lancs.

Buff Plymouth Rock—Mr. & Mrs. R. M. Ault, Alvaston, Derby.

White Plymouth Rock—K. R. Dowrick, St. Austell, Cornwall.

Partridge Plymouth Rock—N. G. Hughes, Leicester.

Breed	Plymouth Rock
Type	Miniature
Classification	Heavy
Origin	U.S.A.
Colour variations	Barred, partridge, buff, white, black, columbian
Egg colour	Cream
Specifications	Stocky, well-built bird, single comb, red ear lobes, yellow legs.
Faults to avoid	Incorrect leg colour or comb.

White Rosecomb—D. C. Mann, Brighouse, Lancs.

Blue Rosecomb—Mrs. B. Dent, Kendal, Cumbria.

Black Rosecomb—Mrs. B. Dent, Kendal, Cumbria.

Breed	Rosecomb
Type	Bantam
Origin	Asia
Colour variations	Black, white, blue
Egg colour	Cream
Specifications	Rose comb, largish white ear lobes. Legs black with black colour, white with white, slate with blue. Male is a graceful bird with long sickle tail.
Faults to avoid	Incorrect comb, lack of white ear lobes.

Scots Dumpy—M. A. Hatcher, Newbury, Berks.

Scots Dumpy—S. Holmes, Newbury, Berks.

Breed	Scots Dumpy
Type	Miniature
Classification	Light
Origin	Scotland
Colour variations	Black, cuckoo
Egg colour	Cream
Specifications	Very short legs, long back with a waddling walk. Single comb, red ear lobes. Tail should be full and flowing.
Faults to avoid	Long-leggedness, five toes.

Scots Grey—R. A. Axman, Dereham, Norfolk.

Scots Grey—R. A. Axman, Dereham, Norfolk.

Breed	Scots Grey
Type	Miniature
Classification	Light
Origin	Scotland
Colour variations	Cuckoo
Egg colour	Cream
Specifications	Single comb, white mottled legs, fine, compact, smart bird, with well-defined markings. Red ear lobes.
Faults to avoid	Heaviness, incorrect comb.

Gold Sebright—C. A. Parker, Preston, Lancs.

Silver Sebright—C. A. Parker, Preston, Lancs.

Breed	Sebright
Type	Bantam
Origin	England
Colour variations	Gold, silver
Egg colour	Cream
Specifications	Pert with tail carried high and prominent breast. Legs slate, rose comb, mulberry colour preferred. Correct lacing most important.
	Citron colour has been obtained by crossing gold and silver.
Faults to avoid	Single comb, fuzzy lacing.

Light Sussex—Mrs. B. Daniel, Otley, Yorks.

Buff Sussex—Mr. & Mrs. F. Cherry, Great Missenden, Bucks.

Silver Sussex—M. Jolley, Sheffield, South Yorks.

Speckled Sussex—Mrs. J. M. Davy, Warwick.

Breed	Sussex
Type	Miniature
Classification	Heavy
Origin	Sussex, England
Colour variations	Light, speckled, buff, white, silver, brown, red
Egg colour	Cream to light brown
Specifications	Single comb, red ear lobes, white legs. Solid carriage. Good egg layer.
Faults to avoid	Legginess, yellow legs, too lightweight, lack of black hackle.

White Transylvanian Naked Neck—D. A. Scrivener.

White Transylvanian Naked Neck—D. A. Scrivener.

Breed	Transylvanian Naked Neck (Turken)
Type	Miniature
Classification	Heavy
Origin	Hungary
Colour variations	White, black, blue, cuckoo, buff, red
Egg colour	Cream
Specifications	Single comb, featherless neck to crop, small cap of feathers surrounding comb. Supposed to have half the number of feathers as chickens of comparable size. Naked areas bright red.
Faults to avoid	Any feathers on neck.

Welsummer—D. Arnold, Bristol.

Welsummer—E. J. Bartlett, Frome, Somerset.

Breed	Welsummer
Type	Miniature
Classification	Light
Origin	Holland
Colour variations	Golden brown
Egg colour	Dark brown
Specifications	Single comb, red ear lobes, yellow legs. Note that the cock and hen have different markings. Compact bird.
	Good egg layer.
Faults to avoid	White legs, too much red in feathers.

Buff-laced Wyandotte—
R. & E. Shaw, Dewsbury, West Yorks.

Blue-laced Wyandotte—
Mr. & Mrs. F. Cherry, Great Missenden, Bucks.

Silver-laced Wyandotte—
O. Storer, Stratford-upon-Avon, Warks.

Gold-laced Wyandotte—
J. J. Holton, Shipston-on-Stour, Warks.

Breed	Wyandotte
Type	Miniature
Classification	Heavy
Origin	U.S.A.
Colour variations	Black, blue, blue-laced, buff, buff-laced, columbian, cuckoo, gold-laced, mottled, silver-laced, partridge, red, silver-pencilled, white
Egg colour	Cream
Specifications	A sturdy breed. Rose comb, red ear lobes, yellow legs. Good egg layer.
Faults to avoid	Over sized rose comb, incorrect leg colour, over size.

Columbian Wyandotte—R. Baker, Sheffield, South Yorks.

Partridge Wyandotte—G. C. Marston, Wetherby, Yorks.

Partridge Wyandotte—P. Kerfoot, Preston, Lancs.

Silver Pencilled Wyandotte—Dr. W. C. Carefoot, Preston, Lancs.

Silver Pencilled Wyandotte—G. C. Marston, Wetherby, Yorks.

Feather variations and markings as comforming to Poultry Club standards

Neck hackle
of Black-Red
Old English Game cock

Neck hackle
of Partridge
Wyandotte cock

Body feather
of Partridge
Wyandotte hen

Neck hackle
of Gold-Laced
Wyandotte cock

Body feather
of Gold-Laced
Wyandotte hen

Neck hackle of
Light Sussex cock

Neck hackle of
Speckled Sussex cock

Body feather of
Speckled Sussex hen

Neck hackle of
Blue Andalusian cock

Body feather of
Blue Andalusian hen

Body feather
of Blue
Belgian Barbu d'Anvers

Body feather of
Rhode Island Red hen

Neck hackle
of Ancona cock

Body feather
of Ancona hen

Neck hackle
of Buff
Plymouth Rock cock

Body feather
of Buff
Plymouth Rock hen

Neck hackle
of Barred
Plymouth Rock hen

Body feather
of Barred
Plymouth Rock hen

Neck hackle
of Marans cock

Body feather
of Marans hen

Body feather of
Indian Game hen

Body feather
of Silver-Laced
Wyandotte hen

Body feather
of Australorp hen

Body feather of
White Silkie hen

Body feather
of Silver-Grey
Dorking hen

Body feather of
Brown Leghorn hen

Body feather
of Gold Duckwing
Old Dutch

Egg colour variation

Dark Brown

Light Brown

Blue/Green

Cream

White

47

The different types of comb

LARGE SINGLE COMB
e.g. Ancona

WALNUT COMB
e.g. Malay

TRIPLE OR PEA COMB
e.g. Brahma

FOLDED SINGLE
COMB
e.g. Leghorn (female)

ROSE COMB
e.g. Sebright

MULBERRY COMB
e.g. Silkie

HORN COMB

SMALL SINGLE COMB
e.g. Old English Game